Abdelhafid Mimouni

Inflammatory Cascade and Oxidative Stress

Abdelhafid Mimouni

Inflammatory Cascade and Oxidative Stress

ScienciaScripts

Imprint

Cover image: www.ingimage.com

This book is a translation from the original published under ISBN 978-620-6-72992-1.

Publisher:
Sciencia Scripts
is a trademark of
Dodo Books Indian Ocean Ltd. and OmniScriptum S.R.L publishing group

120 High Road, East Finchley, London, N2 9ED, United Kingdom
Str. Armeneasca 28/1, office 1, Chisinau MD-2012, Republic of Moldova, Europe
Managing Directors: Ieva Konstantinova, Victoria Ursu
info@omniscriptum.com

Printed at: see last page
ISBN: 978-620-3-32515-7

Inflammatory cascade and oxidative stress

Author: A freelance writer and researcher specialising in bioinorganic chemistry, Dr Mimouni is a recognised expert in the synthesis and characterisation of macromolecules. After obtaining his PhD in Chemistry from the University of Paris XII in 1997, he completed a Diplôme d'Études Approfondies in Bioinorganic Systems at the University of Paris XI in 1993. He also obtained his Bachelor's and Master's degrees in Chemistry from the same university.

Summary: This book explores the complex interaction between **oxidative stress**, **arachidonic acid** and **inflammatory cytokines** in various chronic inflammatory diseases. It describes how the production of **ROS** (reactive oxygen species) and the release of arachidonic acid, activated by complex biochemical processes, trigger an inflammatory cascade, amplified by the production of mediators such as **prostaglandins** and **leukotrienes**. The roles of essential metals such as **iron**, **copper** and **zinc** are also discussed, highlighting their involvement in controlling oxidative stress and inflammatory responses. Using examples from diseases such as **rheumatoid arthritis**, **asthma** and **cardiovascular disease**, the book proposes therapeutic approaches, including **antioxidants** and **IL-6 inhibitors**, to reduce inflammation. Finally, it outlines the prospects for future research and innovative therapeutic strategies in the treatment of chronic inflammatory diseases.

Book outline :

Introduction

Inflammation is an essential defence mechanism in the body, a complex process that fights infection, injury and irritation. Although necessary to maintain the integrity of our immune system, inflammation can become problematic when it becomes chronic or excessive. This phenomenon is at the root of many degenerative, cardiovascular, neurological and autoimmune diseases. Understanding the mechanisms that regulate inflammation is therefore crucial to the development of effective therapeutic strategies.

Among the many biochemical players involved in inflammation, two major phenomena stand out: oxidative stress and the metabolism of polyunsaturated fatty acids, in particular arachidonic acid. Arachidonic acid, released by the action of phospholipase A2 following various inflammatory stimuli, is metabolised into a series of mediators, notably prostaglandins and leukotrienes, which amplify the inflammatory response. However, this biochemical cascade is not isolated. The accumulation of free radicals, responsible for oxidative stress, interacts directly with these processes, exacerbating inflammation and creating a vicious circle of cellular and tissue damage.

But beyond these well-known biological processes, another often overlooked factor plays a decisive role in regulating inflammation: the biochemistry of inorganic elements. Metals such as iron, copper, zinc and manganese play a key role in cellular metabolism and the regulation of oxidative stress. These metal ions are essential cofactors for numerous enzymes, such as superoxide dismutases, cyclooxygenases and lipoxygenases, which play an active role in the production of inflammatory mediators. Their disruption, through excess or deficiency, can lead to an imbalance in cellular metabolism and influence the intensity and duration of the inflammatory response.

This book explores the links between these processes: oxidative stress, arachidonic acid, inflammatory cytokines such as IL-6, and the involvement of bioinorganic elements in these dynamics. By detailing the underlying biochemical mechanisms and illustrating these concepts with concrete examples from various pathologies, this book aims to provide a more complete understanding of the molecular events underlying inflammation. It also offers food for thought on possible therapeutic strategies, highlighting the importance of an integrated approach that takes into account organic mediators as well as metals and antioxidants.

Through this work, we hope not only to provide a synthesis of current knowledge, but also to pave the way for new therapeutic and research avenues in the field of inflammatory biology. This detailed exploration of the biochemical processes and molecular mechanisms involved in inflammation will provide a better understanding of the subtleties of oxidative stress, fatty acid metabolism and the regulation of the immune response, while highlighting the importance of bioinorganics as a key to this understanding.

Chapter 1: Introduction to oxidative stress and inflammation

1.1 What is oxidative stress?

Oxidative stress is a pathophysiological condition in which the balance between the production of reactive oxygen species (ROS) and the body's ability to neutralise these reactive molecules is disturbed. ROS are unstable molecules that contain unpaired oxygen atoms, making them particularly reactive. The main forms of ROS include free radicals such as superoxide (O_2-$^-$), hydrogen peroxide (H_2O_2) and hydroxyl radicals (OH-).

Normally, the body has defence mechanisms, mainly in the form of enzymatic and non-enzymatic antioxidants, to neutralise these reactive species and limit their ability to damage cells. Antioxidant enzymes include **superoxide dismutase (SOD)**, **catalase (CAT)** and **glutathione peroxidase (GPx)**, which are responsible for detoxifying free radicals and hydrogen peroxide. However, in the presence of excess ROS or insufficient defence mechanisms, cells are subjected to oxidative damage, in particular damage to DNA, proteins and lipids. These lesions can lead to cell dysfunction, genetic mutations and, in the long term, chronic pathologies.

The causes of oxidative stress can be diverse, including environmental factors (pollution, radiation, smoking), pathophysiological conditions (obesity, diabetes, cardiovascular disease), infections and even the normal process of energy production in cells. It is also important to note that oxidative stress plays a central role in cellular ageing and in many inflammatory pathologies, where ROS exacerbate the immune response.

1.2 What is inflammation?

Inflammation is a complex and dynamic biological process that represents the immune system's response to physical, chemical or biological aggression. It is the body's defence mechanism for eliminating pathogens, repairing damaged tissue and restoring homeostasis. Acute inflammation is a protective response, characterised by the classic signs of redness, heat, pain, swelling and loss of function. This process involves the rapid and coordinated activation of immune system cells such as macrophages, neutrophils and lymphocytes, as well as the production of inflammatory mediators such as cytokines and prostaglandins.

However, when inflammation becomes chronic, it loses its protective character and can become a risk factor for various diseases. Chronic inflammation is associated with complex diseases, including cardiovascular disease, type 2 diabetes, arthritis, neurodegenerative diseases and certain cancers. This prolonged inflammation may be caused by continuous exposure to inflammatory agents, deregulation of the immune response or persistent oxidative stress.

A central aspect of inflammation is the production of **cytokines**, such as **IL-6**, **TNF-α** and **IL-1β**, which orchestrate the immune response and coordinate the activation of inflammatory cells. These mediators are responsible for amplifying the inflammatory response, vasodilatation and increased permeability of blood vessels, allowing immune cells to migrate to the site of infection or injury.

1.3 Interaction between oxidative stress and inflammation

Oxidative stress and inflammation are closely linked and reinforce each other in many pathologies. In response to an inflammatory agent or injury, immune cells such as macrophages and neutrophils produce ROS. These ROS serve not only to destroy pathogens, but also to modulate the inflammatory response . However, when ROS production is excessive or uncontrolled, they can induce tissue damage and amplify the inflammatory cascade.

On the other hand, inflammation itself contributes to increased oxidative stress. Pro-inflammatory cytokines, such as IL-6, TNF-α and IL-1β, stimulate the production of ROS by immune system cells, thereby reinforcing inflammation. This two-way interaction between oxidative stress and inflammation creates a vicious circle in which one exacerbates the other, leading to chronic inflammation and tissue damage.

Arachidonic acid (AA), a polyunsaturated fatty acid, also plays a central role in this interaction. Released from cell membranes by the action of **phospholipase A2** in response to inflammatory stimuli, arachidonic acid is then metabolised into several classes of pro-inflammatory mediators, such as **prostaglandins**, **leukotrienes** and **thromboxanes**. These molecules act to amplify the inflammatory response by altering vascular permeability, recruiting immune cells and inducing pain.

The production of these inflammatory mediators is strongly influenced by ROS. Indeed, oxidative stress can activate key enzymes in arachidonic acid metabolism, such as **cyclooxygenases (COX)** and **lipoxygenases (LOX)**, which are involved in the production of prostaglandins and leukotrienes respectively. In this way, reactive oxygen species not only aggravate inflammation, but also participate in the production of new inflammatory mediators, creating a negative feedback loop.

In conclusion, oxidative stress and inflammation are two interdependent biological phenomena that reinforce each other in the context of numerous pathologies. Their interaction is mediated by complex biochemical processes involving the production of ROS, the release of arachidonic acid and the regulation of inflammatory cytokines such as IL-6. A better understanding of these mechanisms will enable us to develop more effective therapeutic strategies for managing chronic inflammation-related diseases.

Chapter 2: Arachidonic acid: metabolism and role in inflammation

2.1 Release of arachidonic acid

Arachidonic acid (AA) is a 20-carbon polyunsaturated fatty acid found mainly in the phospholipids of cell membranes, and is a key source of lipid mediators involved in the inflammatory response. One of the first steps in AA metabolism is its release from phospholipid membranes through the action of the enzyme **phospholipase A2 (PLA2)**. This enzyme, activated by inflammatory stimuli (infections, lesions, activation of the immune system), hydrolyses membrane phospholipids, releasing arachidonic acid into the cell's cytosol.

Once released, arachidonic acid can follow different metabolic pathways which determine the nature of the inflammatory mediators produced, and consequently the intensity and duration of the inflammatory response.

2.2 Metabolic pathways of arachidonic acid

2.2.1 Cyclooxygenase (COX) pathway

The **cyclooxygenase pathway** is one of the main metabolic pathways of arachidonic acid. This pathway leads to the formation of **prostaglandins**, **thromboxanes** and **prostacyclins**, which are powerful lipid mediators involved in the regulation of inflammation and coagulation.

- **Prostaglandins (PGs)**: COX catalyses the conversion of arachidonic acid into prostaglandin H2 (PGH2), which is then converted into various prostaglandins, such as **PGE2**, **PGD2** and **PGF2α**. These prostaglandins are involved in vasodilation, increased vascular permeability, and stimulation of pain receptors, contributing to the classic symptoms of acute inflammation: redness, heat and pain. PGE2, in particular, plays a key role in sensitising pain receptors and increasing body temperature during fever.
- **Thromboxanes (TXs)**: COX can also produce thromboxanes, notably **TXA2**, which plays a major role in platelet aggregation and vasoconstriction. This mechanism is particularly important in inflammatory processes associated with vascular lesions, as it contributes to haemostasis and clot formation.
- **Prostacyclins (PGI2)**: Prostacyclins are produced mainly in endothelial cells. Their effects are opposite to those of thromboxanes: they induce vasodilatation and inhibit platelet aggregation. On the other hand, their role in inflammation is more moderate, but important for maintaining the balance between pro- and anti-inflammatory processes.

Cyclooxygenase inhibitors, such as **non-steroidal anti-inflammatory drugs (NSAIDs)**, work by blocking COX activity, thereby reducing the production of these inflammatory mediators and providing relief in acute and chronic inflammatory conditions.

2.2.2 Lipoxygenase (LOX) pathway

Another key pathway in arachidonic acid metabolism is the **lipoxygenase (LOX) pathway**. This pathway leads to the formation of **leukotrienes**, **lipoxins** and other lipid mediators.

- **Leukotrienes (LTs)**: Leukotrienes, such as **LTB4**, **LTC4**, **LTD4** and **LTE4**, are powerful mediators involved in inflammation and modulation of immune responses. LTB4, for example, is a potent chemotactic for neutrophils and plays an important role in the infiltration of leukocytes into inflamed tissues. Leukotrienes, in particular LTC4, LTD4 and LTE4, are also responsible for bronchoconstriction and increased vascular permeability.
- **Lipoxins (LX)**: Unlike leukotrienes, lipoxins are considered to be natural **resolvins**. They have anti-inflammatory properties, helping to limit the duration of the inflammatory response and promote the resolution of inflammation. Lipoxins also interact with specific receptors to regulate immune cell migration and tissue repair.

Lipoxygenase inhibitors can reduce the production of leukotrienes and can be used in inflammatory conditions where these mediators are involved, such as asthma or allergies.

2.2.3 Cytochrome P450 (CYP450) pathway

Another lesser-known route of arachidonic acid metabolism is via the **cytochrome P450** (CYP450) enzymes, which generate a range of products including **epoxides** and **resolvins**.

- **Arachidonic acid epoxides**: Epoxides such as **EETs (arachidonic acid epoxides**) have vasodilatory and anti-inflammatory properties. These metabolites play a role in regulating endothelial function, and can modulate the inflammatory response by inhibiting the production of pro-inflammatory cytokines.
- **Resolvins**: Formed from epoxides or other AA metabolites, resolvins play a crucial role in resolving inflammation, by activating anti-inflammatory pathways that promote resolution of the immune response and limit tissue damage.

These pathways are undergoing rapid research development, as they offer interesting therapeutic potential for chronic inflammatory diseases where resolving inflammation is a key objective.

2.3 Examples of conditions in which arachidonic acid plays a central role

Arachidonic acid and its metabolites play a key role in a multitude of inflammatory and pathological conditions. Some of the most common examples include:

- **Asthma**: In asthma, arachidonic acid is a major player in the inflammatory response of the airways. Leukotrienes, produced via the LOX pathway, induce severe bronchoconstriction, leading to the characteristic symptoms of asthma, such as cough, dyspnoea and asthma attacks. Leukotriene inhibitors, such as **montelukast**, are used to control this inflammatory response.
- **Cardiovascular disease**: Chronic inflammation plays a central role in atherosclerosis, an inflammatory disease of the arteries. Arachidonic acid, via the COX pathway, contributes to the formation of prostaglandins and thromboxanes, which promote vascular inflammation and platelet aggregation. NSAIDs are commonly used to reduce the inflammation associated with these conditions.
- **Rheumatoid arthritis**: In inflammatory arthritis, arachidonic acid is involved in the production of prostaglandins and leukotrienes, which increase joint pain and inflammation. Treatments targeting COX and LOX, such as NSAIDs or specific COX-2 inhibitors, are used to reduce these symptoms.

Conclusion

Once arachidonic acid is released from cell membranes, it follows various metabolic pathways that produce potent mediators such as prostaglandins, leukotrienes and resolvins. These molecules play a key role in regulating inflammation and are implicated in many acute and chronic inflammatory pathologies. The detailed understanding of these pathways and their modulation opens up therapeutic prospects for treating inflammatory diseases and reducing the harmful effects of an excessive inflammatory response.

Chapter 3: Oxidative stress: production and role of ROS

3.1 ROS production in response to inflammation

Oxidative stress is a biochemical phenomenon resulting from the excessive production of reactive oxygen species (ROS), molecules containing highly reactive oxygen, such as free radicals (superoxide, peroxide, etc.). These ROS are generated as part of pathophysiological processes, including inflammation, which is the body's defence response to infection, injury or irritants.

ROS are mainly produced in the **mitochondria**, where the electron transport chain plays a major role in ATP production. In the event of inflammation, this production can be exacerbated by the activation of various enzymes, such as :

1. **NADPH oxidase**: This enzyme is activated by receptors on the surface of immune cells, particularly neutrophils and macrophages. It catalyses the reduction of molecular oxygen (O_2) to superoxide (O_2-$^-$), a highly reactive free radical. This mechanism is crucial in the immune response, where free radicals are used to eliminate pathogens at the inflammatory site (Babior, 2004).
2. **Xanthine oxidase**: Another source of ROS production in inflammation is the enzyme **xanthine oxidase**, which is involved

in the conversion of xanthine to uric acid, releasing superoxide in the process. This pathway is particularly active in cases of metabolic stress or hypoxia (Shin et al., 2009).

3. **Mitochondria**: During cellular dysfunction or stress, mitochondria can produce ROS as a result of electron leakage from the respiratory chain, thereby increasing superoxide production. This mechanism is intensified during inflammatory processes or infections, particularly in immune cells (Turrens, 2003).

3.2 Impact of ROS on cells

ROS are highly reactive molecules that can cause major **cellular damage**, especially when produced in excess during prolonged inflammation. This damage can affect :

1. **Proteins**: ROS can modify amino acid residues in proteins, causing their oxidation. This disrupts their structure and function, which can lead to premature degradation or dysfunction of proteins, affecting processes such as cell signalling or stress response (Ghezzi et al., 2005).
2. **Lipids**: Oxidation of lipids, particularly **membrane lipids**, leads to **lipoperoxidation**, a phenomenon in which free radicals attack cell membranes, altering their fluidity and increasing their

permeability. This can lead to cell membrane rupture, impaired membrane function and, in extreme cases, cell death by necrosis or apoptosis (Frei et al., 1990).

3. **DNA**: ROS can cause DNA damage, such as strand breaks or genetic mutations. This damage can affect the stability of the genome, and if not properly repaired, can contribute to the development of serious pathologies such as cancer (Cadet et al., 2003).

The damage accumulated in cells and tissues as a result of excess ROS contributes to the pathogenesis of various chronic diseases, including cardiovascular disease, neurological disorders and cancer.

3.3 Oxidative stress and inflammatory signalling pathways

Oxidative stress, via ROS, acts as a powerful activator of various **intracellular signalling pathways** that regulate inflammation. Two major pathways, **NF-κB** and **MAPK**, are particularly involved in the ROS-induced inflammatory response.

3.3.1 NF-κB pathway

NF-κB (Nuclear Factor kappa B) is a key transcription factor in the regulation of the inflammatory response. In the presence of ROS, this pathway is activated by the **phosphorylation** of NF-κB inhibitory proteins (IκB), allowing the release and translocation of the NF-κB complex into the nucleus. Once in the nucleus, NF-κB activates the transcription of inflammatory genes that code for **cytokines** (such as TNF-α, IL-1β, IL-6), **prostaglandins** and other pro-inflammatory mediators (Barkin et al., 2001).

Thus, oxidative stress, by activating NF-κB, amplifies inflammation and allows a prolonged response to injury and infection. This activation can become pathological, as in **autoimmune** or **chronic inflammatory diseases** (Baeuerle and Baltimore, 1996).

3.3.2 MAPK pathway

Mitogen-activated protein kinases (MAPKs) comprise several subfamilies, including p38, JNK (c-Jun N-terminal kinase) and ERK (Extracellular signal-regulated kinase). These kinases play a crucial role in regulating inflammation in response to oxidative stress. In the presence of ROS, the MAPK pathway is activated by **successive phosphorylations**, leading to the translocation and activation of transcription factors such as **AP-1** (Activator Protein 1), which promotes the production of pro-inflammatory mediators (Lee et al., 2017).

ROS also regulate the production of cytokines, catabolic enzymes and other molecules involved in tissue degradation, thereby contributing to chronic inflammation. The MAPK pathway is notably involved in acute and chronic inflammatory conditions, such as infections, arthritis and cardiovascular disease (Yang et al., 2003).

3.4 Examples of inflammation where oxidative stress is key

Some chronic inflammatory diseases are particularly sensitive to the effects of oxidative stress. Here are some key examples where ROS play a central role:

1. **Crohn's disease**: This inflammatory bowel disease is characterised by chronic inflammation of the intestinal mucosa. ROS, produced by neutrophils and macrophages, are involved in the breakdown of intestinal tissue. Excessive activation of the NF-κB pathway in intestinal cells can lead to chronic inflammation, exacerbated by oxidative stress (Peyrin-Biroulet et al., 2010).
2. **Rheumatoid arthritis**: In arthritis, joint inflammation is mediated by ROS that activate the NF-κB and MAPK pathways. This induces the production of pro-inflammatory cytokines such as **TNF-α** and **IL-1**, exacerbating joint pain and joint tissue damage (Hunter et al., 2001).
3. **Bacterial infections**: ROS also play a role in host defence against bacterial infections. However, when ROS production becomes excessive, it can lead to collateral damage to neighbouring tissues.

In certain chronic infections, such as tuberculosis or persistent Staphylococcus aureus infections, ROS can increase pathogenicity and contribute to chronic inflammation (Becker et al., 2001).

4. **Cardiovascular disease**: Oxidative stress contributes to atherosclerosis by oxidising low-density lipoprotein (LDL) and activating inflammatory pathways. The accumulation of ROS in endothelial cells provokes an inflammatory response, promoting the formation of atherosclerotic plaques (Madamanchi et al., 2005).

Conclusion

Oxidative stress, induced by ROS, plays a central role in regulating inflammation. While beneficial in the early stages of immune defence, excessive production of ROS can lead to cellular and tissue damage, exacerbating inflammatory responses. The interaction between oxidative stress and inflammatory signalling pathways, such as NF-κB and MAPK, is a key mechanism in the development and chronicisation of many inflammatory diseases.

Chapter 4: Inflammatory cytokines: the role of IL-6

4.1 Role of IL-6 in inflammation

Interleukin-6 (**IL-6**) is a multifunctional cytokine produced by a wide range of cells, including immune cells, endothelial cells and fibroblasts. It plays a central role in the regulation of inflammation and the immune response. IL-6 is activated in many inflammatory processes, where it contributes to the activation of immune cells and the control of acute and chronic responses.

IL-6 activation is regulated by several factors, among which reactive oxygen species (**ROS**) play a key role. In response to inflammatory stimuli, such as infection or tissue injury, immune cells generate ROS that activate intracellular signalling pathways, leading to the production of IL-6.

IL-6 is also regulated by **arachidonic acid** (AA) **receptors**, in particular by the release of AA from cell membranes through the action of **phospholipase A2**. Activated AA directly influences IL-6 production by stimulating inflammatory signalling pathways. These interactions between ROS, AA and IL-6 form a regulatory network that amplifies and prolongs the inflammatory response.

4.2 Molecular mechanisms leading to IL-6 production

Activation of IL-6 production results from a cascade of complex biochemical reactions, involving several **transcription factors** and **kinases**, which are largely regulated by ROS. The main mechanism of IL-6 activation is mediated **by NF-κB**, a key inflammatory signalling pathway in defence and inflammation processes.

4.2.1 NF-κB activation and IL-6 production

The transcription factor **NF-κB** is one of the main players in the regulation of the inflammatory response. In the presence of oxidative stress, **ROS** act on NF-κB inhibitory proteins (IκB), phosphorylating them and causing their degradation. This degradation enables the NF-κB complex to be translocated to the nucleus, where it binds to specific response elements in the promoter of target genes, including that of IL-6. Once activated, **NF-κB** induces the expression of pro-inflammatory genes, including IL-6 (Baeuerle & Baltimore, 1996).

4.2.2 MAPK pathway and interaction with NF-κB

Alongside NF-κB, **mitogen-activated protein kinases (MAPKs)** such as **ERK**, **p38** and **JNK** are activated by ROS. These kinases phosphorylate various proteins, including transcription factors such as **AP-1** and **C/EBP**, which are also involved in the induction of IL-6. This process is enhanced by the activation of **NF-κB**, creating a complex and coordinated network that results in the massive production of IL-6 (Lee et al., 2017).

4.2.3 Interaction with other cytokines and amplification of the response

Once released, IL-6 acts as an amplifier of inflammation. It triggers the production of other cytokines and inflammatory mediators, creating a vicious circle in which the production of IL-6 in turn leads to the production of other inflammatory mediators, such as **TNF-α** and **IL-1β**. For example, **TNF-α**, another major inflammatory cytokine, can promote NF-κB activation, thereby reinforcing IL-6 production in a positive feedback (Schett et al., 2008).

Interactions between **IL-6** and other cytokines influence the diversity of the inflammatory response. For example, under certain conditions, IL-6 can contribute to the transition from acute to chronic inflammation, a process often observed in pathologies such as rheumatoid arthritis or cardiovascular disease (Hunter & Jones, 2015).

4.3 Effects of IL-6 on immune cells

IL-6 exerts its biological effects primarily via its receptor, **IL-6R**, which is expressed on the surface of numerous immune cells, including **T lymphocytes**, **B lymphocytes**, **macrophages** and **neutrophils**. Once bound to its receptor, IL-6 activates various intracellular signalling pathways, such as the **JAK/STAT** (Janus kinase/signal transducer and activator of transcription) pathway. This pathway is particularly important for the differentiation of T lymphocytes, notably in the production of **Th17**, a population of T lymphocytes associated with chronic inflammatory diseases.

1. **Activation of T lymphocytes**: IL-6 plays a key role in the activation and differentiation of T lymphocytes, particularly **Th17 lymphocytes**. Through the production of IL-17, these T cells can promote inflammation in various tissues, particularly in autoimmune and chronic inflammatory diseases such as rheumatoid arthritis (Korn et al., 2007).
2. **Influence on B lymphocytes**: IL-6 is also essential for the differentiation of **B lymphocytes** into antibody-producing **plasma cells.** It is therefore involved in the humoral immune response, which plays a role in infections and certain autoimmune diseases, where excessive antibody production leads to tissue damage.
3. **Effects on macrophages and neutrophils**: IL-6 influences the activation of macrophages and neutrophils, two of the key cells in the acute inflammatory response. It promotes their recruitment to inflammatory sites, as well as the production of cytokines and inflammatory mediators. In chronic conditions, this activation can lead to persistent tissue damage.

4.4 Role of IL-6 in chronic diseases

IL-6 is implicated in the development and progression of many **chronic diseases**, where its role in persistent inflammation contributes to worsening symptoms and tissue damage.

1. **Cardiovascular disease**: IL-6 is a key biomarker in cardiovascular disease, particularly atherosclerosis, where it stimulates atherosclerotic plaque formation and promotes vascular inflammation. It contributes to the transition from acute to chronic inflammation in blood vessels (Harris et al., 2010).
2. **Cancer**: IL-6 is also involved in tumour progression. It promotes tumour cell growth, angiogenesis (formation of new blood vessels) and cell survival by inhibiting apoptotic processes. In certain types of cancer, such as multiple myeloma, IL-6 stimulates the proliferation of cancer cells and contributes to their evasion of immune defence mechanisms (Kumar et al., 2004).
3. **Neuroinflammation**: IL-6 plays an important role in neuroinflammatory disorders, such as Alzheimer's disease and multiple sclerosis. It contributes to brain inflammation, promotes the immune response and may participate in neurodegeneration (Sandler et al., 2014).

Conclusion

IL-6 is a central cytokine in the regulation of inflammation. It is activated by ROS and regulated by arachidonic acid receptors, and participates in a complex signalling network that amplifies the inflammatory response. While its role in activating immune cells and resolving inflammation is essential, its involvement in chronic pathologies such as cardiovascular disease, cancer and neuroinflammation makes it a key factor in the maintenance and progression of chronic inflammation. Controlling IL-6 expression and activity could represent a promising therapeutic target for modulating inflammatory responses in these pathologies.

Chapter 5: Specific examples of inflammatory diseases

5.1 Cardiovascular diseases: Inflammation of the blood vessels

Cardiovascular disease (CVD), and in particular **atherosclerosis**, are inflammatory conditions in which activation of arachidonic acid (AA) metabolism and oxidative stress play a key role in disease progression. Inflammation of the **blood vessels** is at the heart of these pathologies, where an abnormal immune response leads to the formation of atheromatous plaques in the arteries, increasing the risk of stroke, heart attack and other cardiovascular complications.

5.1.1 Activation of arachidonic acid metabolism

Activation of **phospholipase A2** in the endothelial cells of blood vessels releases arachidonic acid from membrane phospholipids. This acid is then metabolised **by the cyclooxygenase (COX)** and **lipoxygenase (LOX)** pathways. The prostaglandins produced by the COX pathway increase inflammation of the arterial walls, promoting vasoconstriction and clot formation. At the same time, leukotrienes from the LOX pathway recruit inflammatory cells such as neutrophils and macrophages, which exacerbate inflammation in the vessel wall.

5.1.2 Oxidative stress and vascular inflammation

Oxidative stress, produced by the activation of **ROS**, also plays a crucial role in vascular inflammation. ROS, generated by endothelial cells and infiltrating immune cells, damage **lipids**, **proteins** and **DNA**, promoting the activation of inflammatory signalling pathways such as **NF-κB** and **MAPK**. These pathways enhance the production of pro-inflammatory cytokines, including **IL-6**, and contribute to the activation of the inflammatory cascade that leads to the progression of atherosclerosis (Yeh et al., 2015).

5.1.3 Examples of cardiovascular diseases

In **atherosclerosis**, an accumulation of lipids and inflammatory cells forms in the walls of the arteries, leading to the formation of plaques. These plaques can rupture and form blood clots, leading to **heart attacks** or **strokes**. Chronic inflammation of the blood vessels, exacerbated by the activation of AA mediators and oxidative stress, is a key factor in the progression of this pathology.

5.2 Rheumatoid arthritis: joint inflammation

Rheumatoid arthritis is an autoimmune disease characterised by chronic inflammation of the joints, particularly the synovial joints. The condition is associated with persistent activation of arachidonic acid metabolism and excessive production of ROS, leading to destruction of cartilage and bone.

5.2.1 Role of arachidonic acid and ROS

In rheumatoid arthritis, activation of **phospholipase A2** releases arachidonic acid, which is metabolised into prostaglandins and leukotrienes. These pro-inflammatory mediators activate **synovial cells**, increasing the production of **IL-6** and other inflammatory cytokines. ROS also play a role in the degradation of cartilage and bone by activating proteolytic enzymes, such as **matrix metalloproteinases (MMPs)**, which degrade the articular extracellular matrix (Harrison et al., 2013).

5.2.2 Inflammation of synoviocytes and disease progression

Synoviocytes, the cells that line the synovial membrane of joints, are activated by arachidonic acid and ROS, promoting the production of IL-6, which stimulates the production of growth factors such as **TNF-α**, and the formation of a chronic inflammatory environment. This inflammatory response persists and leads to the breakdown of joint tissue, with symptoms such as pain, stiffness and loss of joint function.

5.3 Asthma: Inflammation of the respiratory tract

Asthma is a chronic inflammatory disease of the airways, where inflammation is exacerbated by the activation of leukotriene and prostaglandin production. These mediators, produced from arachidonic acid, play a central role in bronchospasm, excessive mucus and airway congestion, leading to typical symptoms such as coughing, wheezing and breathlessness.

5.3.1 The role of leukotrienes and prostaglandins

Leukotrienes, produced by the **lipoxygenase** pathway from arachidonic acid, are powerful mediators of the inflammatory response in asthma. They cause bronchoconstriction, recruitment of inflammatory cells and increased vascular permeability, all of which contribute to airway obstruction. At the same time, **COX-derived** prostaglandins increase vasodilation and vascular permeability, facilitating the infiltration of inflammatory cells into the airways (Chanez et al., 2010).

5.3.2 Oxidative stress and asthma exacerbation

Oxidative stress in the airways plays a key role in the activation of asthmatic inflammation. **ROS** generated by respiratory epithelial cells and inflammatory cells activate signalling pathways such as NF-κB and MAPK, increasing the production of pro-inflammatory cytokines such as IL-6 and worsening inflammation. Increased ROS in the airways can also lead to dysfunction of the antioxidant defence mechanism, contributing to the exacerbation of asthmatic symptoms (Rahman & MacNee, 2000).

5.4 Bacterial and viral infections: Activation of oxidative stress

During **acute immune responses** to bacterial or viral infections, oxidative stress and the production of arachidonic acid play an essential role in the body's defence. Macrophages and other immune cells generate ROS to kill pathogens, but this excess ROS can also lead to local inflammation, which can aggravate tissue damage and prolong the inflammatory response.

5.4.1 Role of ROS and IL-6 in infections

During infection, **ROS** activate inflammatory signalling pathways such as **NF-κB** and **MAPK**, promoting the production of pro-inflammatory cytokines such as IL-6. This cytokine contributes to the mobilisation of immune cells to the site of infection and the amplification of the inflammatory response. However, excessive oxidative stress can also lead to **hyperinflammation**, causing tissue damage and serious complications in infections such as **pneumonia** or **sepsis** (López-Collazo & Hergueta-Redondo, 2018).

5.5 Cancer: chronic inflammation and tumour progression

Chronic inflammation is a key factor in **cancer** progression, with oxidative stress, arachidonic acid and IL-6 playing a central role. The presence of persistent inflammation in the tumour microenvironment promotes the growth, immune evasion and metastasis of cancer cells.

5.5.1 Oxidative stress and cancer

ROS generate mutations in the DNA of cells, increasing the risk of cancerous transformation. ROS also activate pro-inflammatory signalling pathways, such as **NF-κB**, which support tumour cell growth and survival. In cancers such as **colorectal** and **breast cancer**, chronic inflammation linked to oxidative stress and IL-6 activation creates a microenvironment favourable to tumour proliferation and disease progression (Coussens & Werb, 2002).

5.5.2 Role of IL-6 in cancer

IL-6 promotes tumour growth by stimulating processes such as angiogenesis (formation of new blood vessels), cell survival and resistance to apoptosis. In cancers such as multiple myeloma, IL-6 supports tumour cell proliferation and contributes to disease progression by promoting a prolonged inflammatory response (Borrello et al., 2001).

Conclusion

Inflammatory diseases such as cardiovascular disease, rheumatoid arthritis, asthma, infections and cancer are striking examples of the involvement of oxidative stress, arachidonic acid metabolism and IL-6 in the genesis and progression of inflammation. Understanding the molecular mechanisms underlying these processes could provide opportunities for therapeutic interventions targeting these signalling pathways and mitigate the devastating effects of chronic inflammation.

Chapter 6: Therapeutic interventions and prevention

Chronic inflammation, often exacerbated by oxidative stress and disturbances in arachidonic acid metabolism, is a key factor in the development of many inflammatory diseases. In order to modulate this inflammatory response, several therapeutic and preventive approaches have been explored. This chapter focuses on potential interventions aimed at reducing inflammation, the production of inflammatory mediators and oxidative damage.

6.1 Antioxidants : Supplements and reducing oxidative damage

Oxidative stress is caused by an excess of **free radicals** and **reactive oxygen species (ROS)**, which damage cells and tissues, contributing to inflammation and chronic pathologies. **Antioxidants** play an essential role in neutralising these ROS, thereby reducing cellular damage.

6.1.1 Vitamin C and vitamin E

Vitamin C (ascorbic acid) and **vitamin E** (tocopherol) are two of the most studied antioxidants in the context of inflammation. Vitamin C is a water-soluble antioxidant that protects cells against oxidation of lipids, proteins and DNA. It is also involved in the regeneration of vitamin E and in the regulation of certain antioxidant enzymes. As for vitamin E, it is fat-soluble and protects cell membranes from oxidative damage, particularly in lipid-rich tissues such as immune cell membranes.

6.1.2 Polyphenols and other natural antioxidants

Polyphenols, found in fruit, vegetables, tea and red wine, are also powerful antioxidants. These compounds have the ability to neutralise ROS and interfere with inflammatory pathways by modulating transcription factors such as **NF-κB**. Polyphenols, particularly flavonoids, are able to regulate the expression of inflammatory cytokines such as **IL-6** and may thus help to mitigate systemic inflammation (Hernández-Ledesma et al., 2013).

6.1.3 Effectiveness of antioxidant supplements

Although **antioxidant supplements** have shown promising effects in experimental models of inflammatory diseases, the results of clinical trials are often mixed. Studies have suggested that antioxidant intake, while effective in certain conditions, is not a substitute for a balanced lifestyle. What's more, an excess of certain antioxidants can interfere with the body's natural defence mechanisms. Therefore, the most optimal approach remains to promote a diet rich in naturally occurring antioxidants, rather than relying solely on supplements (Halliwell & Gutteridge, 2015).

6.2 COX/LOX inhibitors: Control of inflammation

Inhibition of the **cyclooxygenase (COX)** and **lipoxygenase (LOX)** enzymes, which metabolise arachidonic acid into pro-inflammatory mediators such **as prostaglandins**, **thromboxanes** and **leukotrienes**, is a common strategy for controlling inflammation.

6.2.1 Non-steroidal anti-inflammatory drugs (NSAIDs)

NSAIDs, such as **ibuprofen**, **aspirin** and **naproxen**, inhibit COX activity and thus reduce the production of inflammatory prostaglandins. These drugs are widely used to treat acute and chronic inflammatory conditions, such as **arthritis** or muscle pain. However, their long-term use can be associated with side effects, such as gastric and renal damage, which limits their application in certain populations.

6.2.2 5-lipoxygenase inhibitors

Inhibitors of **5-lipoxygenase** (5-LOX) block the production of leukotrienes, powerful mediators involved in inflammatory conditions such as **asthma** and **hay fever**. Drugs such as **zileuton**, which inhibit this enzyme, have shown benefits in controlling asthma by reducing bronchoconstriction and inflammation of the airways. However, these inhibitors are not without side effects, such as gastrointestinal disorders and liver abnormalities.

6.3 Modulation of IL-6: targeted therapies

IL-6 plays a central role in inflammation and in many chronic diseases. As a result, targeted therapies aimed at inhibiting the action of IL-6 or its receptors have been developed to treat inflammatory pathologies.

6.3.1 IL-6 inhibitors: Tocilizumab

Tocilizumab, a monoclonal inhibitor of IL-6, is used in the treatment of **rheumatoid arthritis** and other chronic inflammatory diseases. It works by binding to the IL-6 receptor, blocking its signalling and reducing the production of pro-inflammatory cytokines. Tocilizumab has shown promising results in treating systemic inflammation and the progression of joint damage in patients with rheumatoid arthritis (Smolen et al., 2016).

6.3.2 Other targeted therapeutic approaches

Similar therapies targeting other cytokines, such as **TNF-α** (TNF inhibitors, e.g. adalimumab), are also used to treat chronic inflammatory diseases such as **ulcerative colitis** and **ankylosing spondylitis**. These therapies have been shown to reduce symptoms and modify disease progression, but they are also associated with risks of immunosuppressive side effects and infections.

6.4 Anti-inflammatory diet: Modulation of the inflammatory response

Diet plays a crucial role in regulating inflammation, and a suitable diet can be an effective way of modulating the inflammatory response over the long term.

6.4.1 A diet rich in omega-3 and low in saturated fatty acids

Omega-3 fatty acids, found in oily fish (such as salmon and mackerel), walnuts and linseed, have well-established anti-inflammatory properties. They act by inhibiting the production of inflammatory mediators derived from arachidonic acid and by promoting the production of **resolvins** and other anti-inflammatory mediators. Increased omega-3 intake has been associated with a reduction in inflammatory symptoms in conditions such as **rheumatoid arthritis** and **asthma**.

On the other hand, a diet rich in **saturated fatty acids** (found in fatty meats, full-fat dairy products and processed foods) can increase the production of arachidonic acid and inflammatory mediators, exacerbate oxidative stress and contribute to chronic inflammation.

6.4.2 Antioxidants in food

A **diet rich in fruit and vegetables**, which provides significant amounts of **vitamin C**, **vitamin E** and **polyphenols**, can also help to reduce inflammation. **Flavonoids**, for example, have proven antioxidant and anti-inflammatory effects, by modifying cell signalling pathways and inhibiting the production of inflammatory cytokines such as **IL-6**.

Conclusion

Therapeutic interventions aimed at reducing inflammation and oxidative stress, such as the use of antioxidants, COX and LOX inhibitors, and IL-6 modulation, represent important strategies for the management of inflammatory diseases. At the same time, a preventive approach including an **anti-inflammatory diet** rich in omega-3 and antioxidants may play a key role in reducing chronic inflammation and preventing long-term inflammatory diseases.

Chapter 7: The role of bioinorganics in inflammation and oxidative stress

1. Introduction: What is bioinorganics and why are they relevant to inflammation?

Inorganic biochemistry focuses on the study of non-organic chemical elements, such as **metal ions**, and their influence on essential biological processes. This discipline explores the roles of these elements in mechanisms such as **enzyme catalysis, redox reactions** (reductions and oxidations) and the **regulation of metabolic pathways**. **Metal ions** act as cofactors in many biochemical reactions, influencing processes crucial to cellular health, including **inflammation** and **oxidative stress**.

Link with inflammation

Metal ions are directly involved in the mechanisms of **oxidative stress**, the process by which reactive oxygen species (**ROS**) are produced. These ROS are essential in triggering acute and chronic inflammation. When these metals are poorly regulated or accumulate, they exacerbate oxidative stress, damage cells and modify inflammatory responses. For example, the **accumulation of iron** in certain pathological conditions can lead to an **excessive inflammatory response**, thus contributing to chronic inflammatory diseases.

2. Metal ions as catalysts in fatty acid metabolism and ROS production

Metal ions, such as iron, copper and zinc, are involved in fundamental processes such as the production of **ROS** and the regulation of **fatty acids**. These elements facilitate the production of **inflammatory metabolites** that modulate inflammation and oxidative stress.

Iron (Fe)

Iron is crucial in the production of **ROS**, in particular via the **Fenton reaction**. In this reaction, ferrous iron (Fe^{2+}) catalyses the conversion of **hydrogen peroxide** (H_2O_2) into **hydroxyl radicals** (OH-), highly reactive chemical species capable of damaging **cell membranes**, **DNA** and **proteins**.

- **Example**: The activation of **NF-κB**, a key transcription factor in the regulation of inflammation, is influenced by ROS. These ROS, in the presence of iron, are able to activate **NF-κB**, triggering the production of inflammatory cytokines and amplifying the inflammatory response. This process is particularly critical in **pathologies linked to iron metabolism**, such as **haemochromatosis**, where excessive iron accumulation can lead to systemic inflammation.

Copper (Cu)

Copper is involved in the regulation of **antioxidant enzymes** such as **superoxide dismutase** (SOD). SOD converts **superoxide radicals** (O_2-) into **hydrogen peroxide** (H_2O_2), thereby reducing oxidative stress. Copper also plays an important role in enzymes such as **cytochrome c oxidase**, involved in **cellular respiration**.

- **Example**: In conditions such as **Wilson's disease**, where copper accumulates in a toxic way, **inflammation** is exacerbated, leading to tissue damage and a prolonged inflammatory response.

Zinc (Zn)

Zinc is an essential element in many biological processes, notably as a cofactor for enzymes regulating fatty acid metabolism and inflammatory cytokines. Zinc interacts with transcription factors such as **NF-κB**, influencing the production of pro-inflammatory cytokines such **as IL-6**.

- **Example**: **Zinc deficiency** is associated with increased **systematic inflammation**, contributing to pathologies such as **rheumatoid arthritis**, where levels of inflammatory cytokines, particularly **IL-6**, are high.

3. Manganese and other trace elements in inflammatory mechanisms

In addition to iron, copper and zinc, other metallic elements such as **manganese (Mn)** also play an important role in regulating inflammation and oxidative stress.

Manganese (Mn)

Manganese is a cofactor of the enzyme **manganese superoxide dismutase** (Mn-SOD), located in the **mitochondria**. This enzyme protects cells against **superoxide radicals** (O_2-) generated during energy processes. Manganese therefore plays a key role in managing oxidative stress.

- **Example**: Manganese is implicated in **neurodegenerative diseases** such as **Parkinson's**, where chronic inflammation of the brain is exacerbated by abnormal levels of manganese, contributing to the progression of the disease.

Other items

Selenium (Se), **calcium** (Ca) and **magnesium** (Mg) are also involved in controlling oxidative stress and regulating the inflammatory response. Selenium, for example, is a key constituent of **glutathione peroxidase**, an antioxidant enzyme that reduces oxidative stress at cellular level.

4. Bioinorganics and the production of inflammatory mediators

Metal ions also influence the production of **inflammatory mediators**, which regulate the duration and intensity of inflammation.

Phospholipase A2 (PLA2)

PLA2 is an enzyme that releases **arachidonic acid** from cell membranes. This arachidonic acid is then converted into inflammatory mediators such as **prostaglandins** and **leukotrienes**, via the **COX** and **LOX** pathways respectively. **magnesium** is an essential cofactor for the activity of certain PLA2 isoforms, influencing the production of these mediators.

Cyclooxygenase (COX) and Lipoxygenase (LOX)

The **COX** and **LOX enzymes** play a central role in the production of prostaglandins and leukotrienes respectively. These mediators are involved in vasodilatation, blood vessel permeability and immune cell infiltration. Copper and zinc regulate these enzymes, and an imbalance in fatty acid metabolism can promote chronic inflammation.

5. Metal complexes in the modulation of IL-6 and other cytokines

Metal ions also influence the production of **inflammatory cytokines** such as **IL-6**, which is a key mediator in the regulation of the immune response.

- **Example**: In diseases such as **rheumatoid arthritis** or **cardiovascular disorders**, **zinc** deficiency can lead to overproduction of inflammatory cytokines, particularly **IL-6**, which contributes to the progression of inflammation.

Metallic inhibitors, such as those targeting metalloproteinases (enzymes that regulate tissue degradation and are involved in inflammation), are showing therapeutic potential in reducing chronic inflammation.

6. Interactions between oxidative stress, inflammation and metals: a feedback loop

Disturbances in **metal** metabolism can lead to overproduction **of ROS**, amplifying inflammation.

- **Example**: In diseases such as **thalassaemia** and **haemochromatosis**, the accumulation of iron generates excessive production of free radicals, exacerbating inflammation and damaging tissues.

7. Examples of pathologies in which bioinorganics are involved

Metal imbalances have important implications for various inflammatory pathologies.

Cardiovascular diseases

Imbalances in metals, such as excess **iron** or **zinc** deficiency, can contribute to vascular inflammation, promoting conditions such as **atherosclerosis**. These imbalances increase the production of ROS and disrupt signalling pathways, exacerbating blood vessel inflammation.

Cancer

Metal ions such as iron and copper are implicated in **tumour** progression by modulating the chronic inflammatory response. High levels of iron can encourage tumour growth, particularly in cancers such as **colorectal** and **breast**, by exacerbating local inflammation and promoting genetic mutations.

Neurodegenerative diseases

Metallic elements, in particular **iron**, **copper** and **zinc**, are associated with **neurodegenerative diseases**, where their accumulation in the brain promotes chronic inflammation and high oxidative stress, contributing to pathologies such as **Parkinson's disease** and **Alzheimer's**.

8. Conclusion: Towards new therapeutic approaches based on bioinorganics

Metal ions play a central role in controlling inflammation and oxidative stress. Appropriate management of these metal levels, whether by regulating mineral intake or targeting **metalloproteinases** and other **metal complexes**, offers interesting therapeutic prospects for treating chronic inflammatory diseases.

Conclusion

The complex interaction between **oxidative stress**, **arachidonic acid** and **inflammatory cytokines** plays a central role in the development and progression of many **inflammatory diseases**. These three elements are intimately linked in the regulation of the inflammatory response, each influencing the other in a vicious circle that amplifies inflammation and tissue damage.

Oxidative stress, through the production of **ROS**, can activate various inflammatory pathways, including the production of pro-inflammatory **cytokines** such as **IL-6**, and modulate processes such as the release of **arachidonic acid**. This acid, once released by the action of **phospholipase A2**, becomes an essential precursor of inflammatory mediators such as **prostaglandins** and **leukotrienes**, which regulate vasodilation, cellular infiltration and other inflammatory responses. In turn, these mediators can amplify oxidative stress, creating a feedback loop that feeds and prolongs inflammation.

Understanding these interconnections offers crucial prospects for the treatment of **chronic inflammatory diseases** such as **rheumatoid arthritis**, **cardiovascular** disease, **asthma** and **cancer**. By targeting key molecules such as **arachidonic acid**, inflammatory **cytokines** (such as **IL-6**) and sources of **oxidative stress**, it becomes possible to develop

more effective therapies that interrupt these inflammatory feedback loops. In addition, the integration of nutritional strategies and antioxidant supplements could play a role in managing oxidative stress levels, thereby helping to limit inflammation.

Practical implications for clinical treatment and future research :

1. **Clinical treatments**: The management of inflammatory diseases could benefit from targeted therapies aimed at specifically inhibiting **cyclooxygenases** (COX), **lipoxygenases** (LOX), or even **IL-6**, while reducing **oxidative stress** levels through the use of **antioxidant supplements** or metalloproteinase inhibitors. These strategies can not only reduce inflammation, but also limit long-term tissue damage.
2. **Future research**: Future research should focus on more targeted approaches to modulating these biochemical processes. Exploring **therapies based on metals** such as copper, zinc and iron, as well as the role of **trace elements** in regulating inflammation, could open up new therapeutic avenues. In addition, in-depth studies on the interaction between **metals** and **inflammatory metabolites** are needed to understand how these elements contribute to the development of chronic diseases and how their modulation could improve treatments.

In short, a better understanding of the interaction between oxidative stress, arachidonic acid and inflammatory cytokines could lead to the development of more personalised, effective and sustainable treatments for a wide range of chronic inflammatory diseases. Continued research in this area is essential to broaden our therapeutic horizons and respond more precisely to patients' needs.

Glossary :

- **Oxidative stress**: Imbalance between the production of reactive oxygen species (ROS) and the body's antioxidant capacity, leading to cell damage.
- **Reactive oxygen species (ROS)**: Molecules containing unpaired oxygen atoms, such as superoxide (O_2-$^-$), hydrogen peroxide (H_2O_2) and hydroxyl radicals (OH-), responsible for oxidative stress.
- **Cytokines**: Proteins secreted by immune cells that regulate the inflammatory response.
- **Arachidonic acid (AA)**: Polyunsaturated fatty acid released from cell membranes and metabolised into inflammatory mediators such as prostaglandins, leukotrienes and thromboxanes.
- **Phospholipase A2**: Enzyme which releases arachidonic acid from cell membrane phospholipids in response to inflammatory signals.
- **Cyclooxygenases (COX)**: Enzymes which convert arachidonic acid into prostaglandins, key mediators of inflammation.
- **Lipoxygenases (LOX)**: Enzymes which metabolise arachidonic acid into leukotrienes, also involved in inflammation.
- **Cytochrome P450 (CYP450)**: Enzyme involved in the alternative metabolism of arachidonic acid, producing epoxides and resolvins.
- **Prostaglandins (PG)**: Lipid drugs involved in vasodilatation and amplification of the inflammatory response.
- **Leukotrienes (LT)**: Mediators involved in bronchoconstriction and

infiltration of immune cells.
- **IL-6 (Interleukin-6)**: Major pro-inflammatory cytokine involved in many inflammatory and autoimmune diseases.
- **NF-κB (Nuclear Factor kappa B)**: Transcription factor crucial in the regulation of pro-inflammatory genes, including IL-6.
- **MAPK (Mitogen-Activated Protein Kinase)**: Intracellular signalling pathway involved in the activation of various cytokines, including IL-6.
- **JAK/STAT**: Intracellular signalling pathway essential for the production of cytokines, including IL-6.
- **Th17**: T lymphocytes involved in chronic inflammation, often regulated by IL-6.
- **Atherosclerosis**: Accumulation of lipid deposits and inflammatory cells in artery walls, leading to plaque formation and vessel obstruction.
- **Antioxidants** : Substances that neutralise free radicals and reactive oxygen species (ROS), thereby reducing cell damage.
- **Tocilizumab**: IL-6 inhibitor used to treat chronic inflammatory diseases such as rheumatoid arthritis.
- **Resolvins**: Anti-inflammatory mediators produced from omega-3 fatty acids, which help to resolve inflammation.
- **Omega-3**: Essential fatty acids found in oily fish and certain plants, which have anti-inflammatory properties.
- **Superoxide dismutase (SOD)**: Enzyme that catalyses the conversion of superoxide radicals (O_2-) into hydrogen peroxide (H_2O_2), an important

antioxidant.

- **Metalloproteinases**: Enzymes that break down components of the extracellular matrix, regulating cell growth and inflammation.

References

- Halliwell, B., & Gutteridge, J. M. (2015). *Free Radicals in Biology and Medicine*. Oxford University Press.

- Hernández-Ledesma, B., et al. (2013). Antioxidant properties of polyphenols in food and human health. *Current Drug Targets, 14*(3), 318-327.

- Smolen, J. S., et al. (2016). Tocilizumab in early rheumatoid arthritis. *New England Journal of Medicine, 373*, 1043-1050.

- Borrello, I., et al. (2001). Interleukin-6 and its receptor as therapeutic targets in the myeloma. *Clinical Cancer Research, 7*(4), 1188-1195.

- Chanez, P., et al. (2010). Role of inflammatory mediators in the pathogenesis of asthma. *American Journal of Respiratory and Critical Care Medicine, 182*(3), 297-303.

- Coussens, L. M., & Werb, Z. (2002). Inflammation and cancer. *Nature, 420*(6917), 860-867.

- Harrison, J. W., et al. (2013). Role of reactive oxygen species in rheumatoid arthritis. *Journal of Immunology Research*, 2013, 937272.

- López-Collazo, E., & Hergueta-Redondo, M. (2018). The role of oxidative stress in sepsis. *Journal of Immunology Research, 2018*, 1079786.

- Yeh, C. H., et al. (2015). Inflammation and atherosclerosis: The role of oxidative stress. *Journal of Cardiovascular Research, 106*(1), 1-11.

- Baeuerle, P. A., & Baltimore, D. (1996). NF-kappa B: ten years after. *Cell, 87*(1), 13-20.
- Harris, T. B., et al. (2010). Inflammatory markers and the risk of cardiovascular disease in the elderly. *The New England Journal of Medicine, 360*(3), 199-207.
- Hunter, C. A., & Jones, S. A. (2015). IL-6 as a keystone cytokine in health and disease. *Nature Immunology, 16*(5), 448-457.
- Korn, T., et al. (2007). IL-6 and TGF-β1: partners in promoting autoimmune inflammation. *Nature Immunology, 8*(9), 929-935.
- Kumar, S., et al. (2004). Role of interleukin-6 in the pathogenesis of multiple myeloma. *Clinical Cancer Research, 10*(3), 1353-1361.
- Lee, J., et al. (2017). ROS generation and the involvement of the MAPK signaling pathway in inflammatory diseases. *Free Radical Biology and Medicine, 112*, 85-92.
- Sandler, N., et al. (2014). Role of IL-6 in neuroinflammation. *Journal of Clinical Investigation, 124*(9), 3910-3918.
- Schett, G., et al. (2008). IL-6 and TNF-alpha in rheumatoid arthritis. *Annals of the Rheumatic Diseases, 67*(5), 575-582.
- Samuelsson, B. (1983). Arachidonic acid metabolism. *The FASEB Journal,* 2(3), 211-219.
- Funk, C. D. (2001). Prostaglandins and leukotrienes: Advances in eicosanoid biology. *Science, 294*(5548), 1871-1875.

- Murphy, R. C., & Walker, M. E. (2009). Lipid mediators in inflammation. *Current Opinion in Lipidology, 20*(3), 194-200.
- FitzGerald, G. A., & Patrono, C. (2001). The Coxibs, selective inhibitors of cyclooxygenase-2. *The New England Journal of Medicine, 345*(6), 433-442.
- Finkel, T., & Holbrook, N. J. (2000). Oxidants, oxidative stress, and the biology of aging. *Nature, 408*(6809), 239-247.
- Medzhitov, R. (2008). Origin and physiological roles of inflammation. *Nature, 454*(7203), 428-435.
- Tewari, M., & Hoh, J. (2012). The role of inflammation in aging and disease. *Biochimica et Biophysica Acta (BBA) - Molecular Basis of Disease, 1822*(1), 1-10.
- Libby, P. (2007). Inflammation in atherosclerosis. *Nature, 420*(6917), 868-874.
- Calder, P. C. (2006). n-3 polyunsaturated fatty acids and inflammation. *The American Journal of Clinical Nutrition, 83*(6), 1505S-1519S.

Printed by Books on Demand GmbH, Norderstedt / Germany